Section 1

Biological molecules

All living things are made up of the same organic molecules, which is compelling evidence for evolution.

Carbohydrates are the main respiratory substrate used in living cells. They are also molecules that have structural roles, such as in cell membranes and cell walls. Lipids are found in cell membranes and can be used as respiratory substrates. Many hormones are lipids. Proteins have many uses in cells, including as enzymes or hormones, in cell membranes and as components of the blood. Nucleic acids act as the genetic code. They carry the information used to synthesise proteins in cells. The same genetic code is found in all living things and in viruses, which is further compelling evidence for evolution. The main component of all cells is water, which is why scientists searching for life elsewhere in the universe look for the presence of liquid water.

Monomers and polymers

Most biological molecules are very large. They are made by joining together many smaller units, called monomers, into long chains. The resulting large molecule is a polymer. The monomers join together by a condensation reaction. When these large biological molecules are digested, they are broken down into smaller components by hydrolysis. Examples of monomers in biology are monosaccharides, amino acids and nucleotides.

1 **What is a monomer? (AO1)** 1 mark

....

....

2 **Give *two* examples of biological polymers. (AO1)** 1 mark

....

....

3 **What is a condensation reaction? (AO1)** 1 mark

....

....

4 **What is a hydrolysis reaction? (AO1)** 1 mark

....

....

Carbohydrates

Carbohydrates are made of carbon, hydrogen and oxygen. Monosaccharides are simple sugars, such as glucose. Glucose is commonly used as a respiratory substrate. Two monosaccharides can join together by a condensation reaction to form disaccharides such as maltose, sucrose and lactose. Polysaccharides are polymers of monosaccharides. These include starch, glycogen and cellulose, which are formed by condensation reactions between α-glucose units, or β-glucose units in the case of cellulose. The structure of these polysaccharides is related to their functions. Starch and glycogen are storage polysaccharides, whereas cellulose has a structural role in plant cell walls.

1 Glucose, galactose and fructose all have the same formula of $C_6H_{12}O_6$ but they are different sugars. Explain how. (AO2) **1 mark**

....

....

2 The diagram shows the disaccharide trehalose. Annotate the diagram to show how it can be hydrolysed into two molecules of glucose. (AO2) **2 marks**

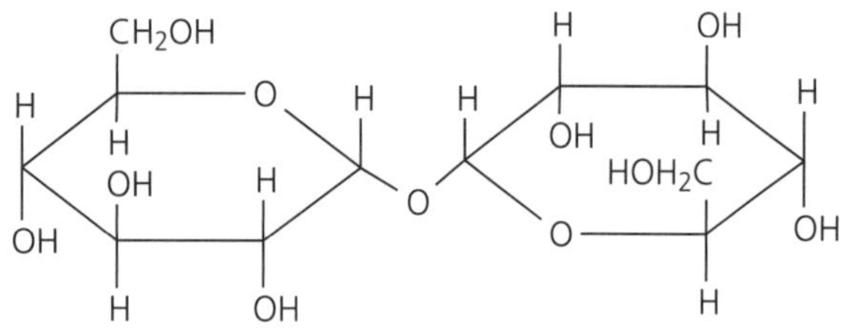

CH_2OH H OH
H O H H H
H OH H
OH H HOH_2C
O
OH O OH
H OH H

3 The diagram shows sophorose. This is an unusual disaccharide which, like trehalose, is made from two α-glucose molecules.

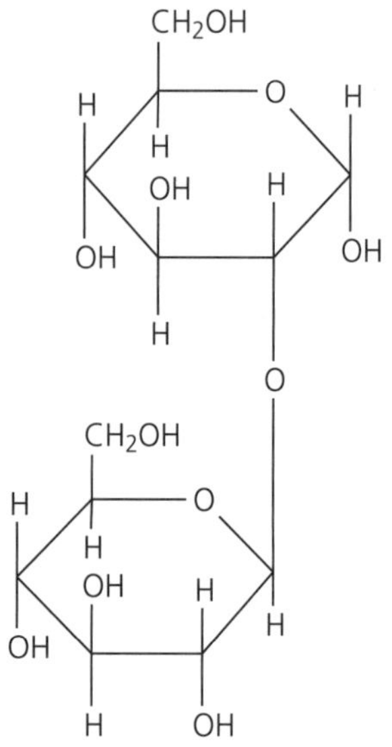

CH_2OH
H O H
H
OH H
OH OH
H
O
CH_2OH
H O
H
OH H
H
OH
H OH

a Draw a circle round the glycosidic bond. (AO1) **1 mark**

b This molecule is a different shape from trehalose. Explain why. (AO2) **1 mark**

....

....

4 The diagram shows α-glucose and β-glucose. Describe the difference between them. (AO1)

1 mark

α-glucose

H O H
HO OH

β-glucose

H O OH
HO H

....

....

5 Complete the table to show the monosaccharides in each of the polysaccharides listed. (AO1)

2 marks

Polysaccharides	Monosaccharides
Glycogen	
Cellulose	
Starch	

6 Explain *two* ways in which the structure of glycogen and starch are related to their function. (AO1)

....

....

....

....

7 Explain *two* ways in which the structure of cellulose is related to its function. (AO1)

....

....

....

....

8 Describe how you could test a solution for:

a starch (AO1)

....

....

....

....

b reducing sugar (AO1)

....

....

....

....

9 **Describe how you could test a solution to show that it contains a non-reducing sugar. (AO1)** 3 marks

Lipids

Triglycerides — commonly called fats and oils — are made of three fatty acids joined to a glycerol molecule. Phospholipids are made from a glycerol molecule joined to two fatty acids, with a phosphate group attached instead of a third fatty acid. These are polar molecules that are important in the structure of cell membranes.

1 **The diagram shows the structure of glycerol and a fatty acid. In the space below, show how these molecules join together to form a triglyceride. Name the bond formed. (AO1)** 3 marks

Glycerol

```
     H
     |
H — C — OH
     |
H — C — OH
     |
H — C — OH
     |
     H
```

Fatty acid

R—COOH

2 **Name the reaction that occurs when a triglyceride is formed from fatty acids and glycerol. (AO1)** 1 mark

3 **Describe the difference between a saturated and an unsaturated fatty acid. (AO1)** 1 mark

4 Describe how you could use the emulsion test to test for the presence of a lipid. (AO1) **3 marks**

....

....

....

....

....

....

5 The diagram shows the structure of a phospholipid.

A

B

C

a Name the parts of the molecule labelled A, B and C. (AO1) **3 marks**

A .... B .... C ....

b A phospholipid molecule is said to be 'polar'. Explain what this means. (AO1) **2 marks**

....

....

....

....

17

Exam-style questions

1 Olestra is a fat substitute. It is made from the disaccharide sucrose, with six to eight fatty acid molecules attached to it.

a i Complete the table to give *two* differences between Olestra and a normal fat. (AO2) **2 marks**

Olestra	Normal fat

ii Circle the name of a bond found in Olestra. (AO2) **1 mark**

glycosidic ester peptide disulfide bridge phosphodiester

iii Explain your answer. (AO2) **1 mark**

b Olestra passes through the gut and into the faeces without being digested. Explain why. (AO2) **2 marks**

c In a study lasting 3 months, 76 volunteers were used.

- The volunteers were divided into two groups, a control group and an experimental group.
- The volunteers in the experimental group consumed between 20 and 40 g of Olestra daily, which was incorporated into foods such as chips, meat pies, milk and biscuits.
- Neither the volunteers nor the scientists analysing the results knew which volunteers were eating Olestra and which were not.
- The food each person ate was monitored closely and the scientists found that total fat consumption was significantly reduced following 3 months on Olestra.

i It was important that neither the volunteers nor the scientists analysing the results knew which volunteers were eating Olestra and which were not. Explain why. (AO3) **1 mark**

ii Suggest how the control group was treated. (AO3) **2 marks**

2 The waterproof covering of the skin is made of lipids. Scientists studying this lipid layer have found that it is made of phospholipids. However, the phospholipids found in the skin's waterproof covering are splayed outwards so that the two 'tails' of each molecule point in opposite directions. These molecules are then stacked on top of each other in an alternating fashion. This gives a tightly packed structure that is more impermeable than a normal phospholipid bilayer.

a What property of phospholipids enables them to form a bilayer in a membrane? (AO1) **2 marks**

b **In a cell-surface membrane, phospholipids form a bilayer. Explain how. (AO1)** **2 marks**

....

....

....

....

c **i** **Give *one* difference between the phospholipids found in a cell-surface membrane and the phospholipids found in the skin's lipid layer. (AO2)** **1 mark**

....

....

ii **Explain how the arrangement of the phospholipids in the skin's lipid layer makes it more impermeable than a normal phospholipid bilayer. (AO2)** **2 marks**

....

....

....

....

Proteins

Proteins are large molecules. They are polymers of amino acids. Their structure can be described in different ways, i.e. their primary, secondary and tertiary structure. Some large proteins have a quaternary structure. The shape of proteins is important for their function.

General properties of proteins

1 **The diagram shows an amino acid.**

```
H     H    O
 \    |   //
  N — C — C
 /    |    \
H     R     OH
```

Label the amino group, carboxylic acid group and the part of the molecule that varies between different amino acids. (AO1) **3 marks**

2 **In the space below, draw a diagram to show how two amino acids join together to form a dipeptide. Name the bond formed and the type of reaction involved. (AO1)** **4 marks**

3 **Complete the table to describe the different levels of structure of proteins and the bond(s) holding the protein in this structure. (AO1)** 6 marks

Structure	Description	Bond(s) holding this structure in place
Primary structure		
Secondary structure		
Tertiary structure		

4 **Some proteins have a quaternary structure. What is a quaternary structure? (AO1)** 2 marks

....

....

....

....

5 **Describe how you could use the biuret test to show that a solution of albumen contains protein. (AO1)** 2 marks

....

....

....

....

Many proteins are enzymes

Enzymes are proteins that lower the activation energy needed for reactions to occur. Each enzyme is specific to its substrate. It has a precise tertiary structure that results in an active site that is complementary to its substrate. They work by the induced-fit model.

The rate of an enzyme-controlled reaction is affected by pH, temperature, substrate concentration, enzyme concentration, competitive and non-competitive inhibitors. Enzymes catalyse a wide range of reactions, both inside and outside cells.

1 **The graph shows energy changes during a reaction.**

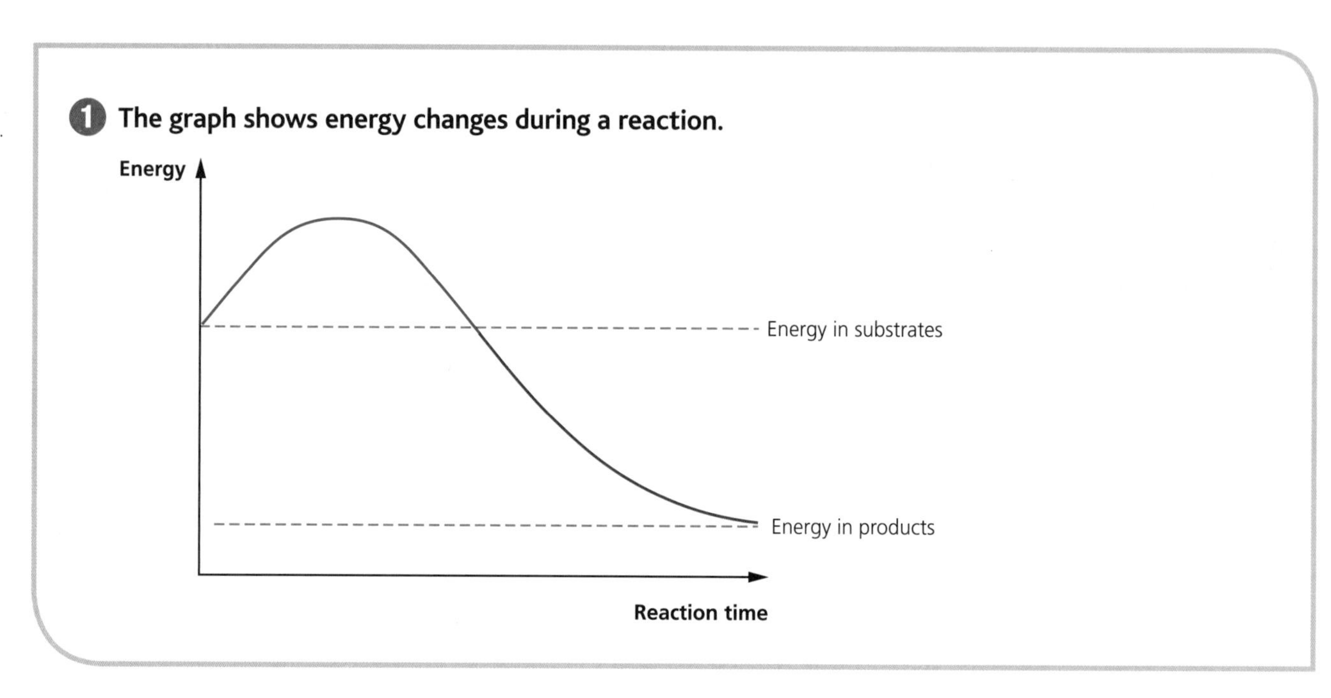

a Add a label to show activation energy. (AO1) **1 mark**

b The graph shows the energy changes when an enzyme is not present. Add a line to show the energy changes when an enzyme is present. (AO1) **2 marks**

2 Describe the induced-fit model of enzyme action. (AO1) **3 marks**

3 Use your knowledge of protein structure to explain why amylase will digest starch but not glycogen. (AO1) **3 marks**

4 The graph shows the effect of temperature on an enzyme-catalysed reaction.

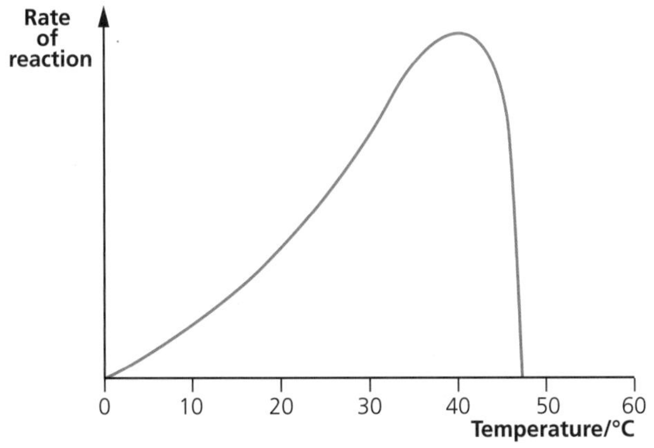

Rate of reaction

0 10 20 30 40 50 60

Temperature/°C

a Explain why the rate of reaction increases between 0°C and 35°C. (AO1) **2 marks**

b **Explain why the rate of reaction decreases rapidly between 40°C and about 48°C. (AO1)** **3 marks**

....

....

....

....

....

....

....

5 **The graph shows the effect of pH on an enzyme-catalysed reaction.**

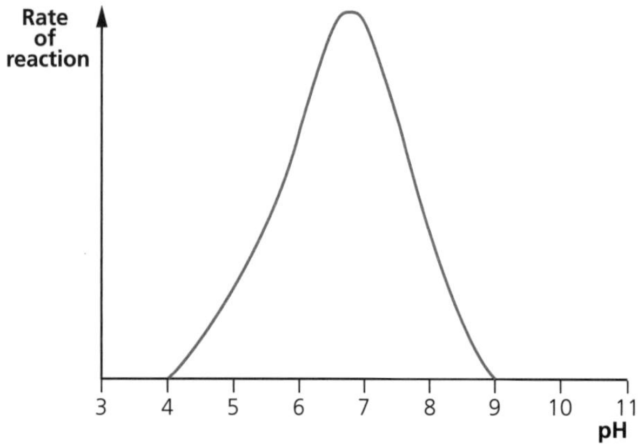

Explain the shape of the graph. (AO1) **3 marks**

....

....

....

....

....

....

....

....

6 **The graph shows the effect of substrate concentration on an enzyme-catalysed reaction.**

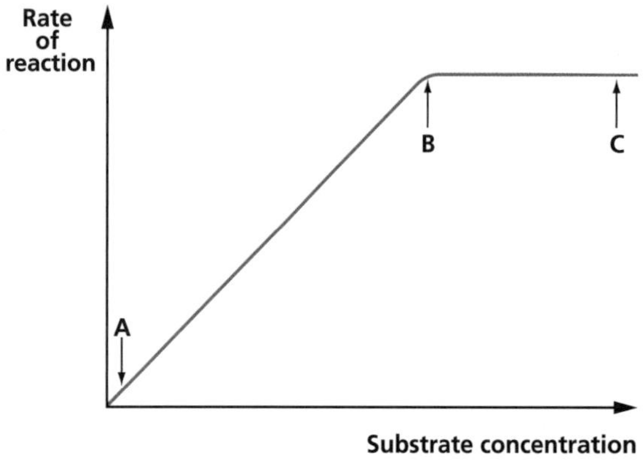

a Describe the shape of the graph. (AO2) **2 marks**

b Explain the shape of the graph between:

i A and B (AO1) **2 marks**

ii B and C (AO1) **1 mark**

7 Describe the effect of enzyme concentration on the rate of an enzyme-catalysed reaction in the presence of excess substrate. (AO1) **1 mark**

8 Explain how a competitive inhibitor inhibits enzyme activity. (AO1) **3 marks**

9 Explain how a non-competitive inhibitor inhibits enzyme activity. (AO1) **3 marks**

10 Adding more substrate reduces the inhibitory effect of a competitive inhibitor. Explain how. (AO1) **2 marks**

11 Adding more substrate does not reduce the effect of a non-competitive inhibitor. Explain why. (AO1) **2 marks**

....

....

....

....

12 Quinacrine mustard is similar in shape to the substrate of the enzyme trypanothione reductase. It inhibits this enzyme, which is found in the parasite *Trypanosoma*, the cause of sleeping sickness.

a Explain how quinacrine mustard inhibits the enzyme. (AO2) **2 marks**

....

....

....

....

b This enzyme is essential in *Trypanosoma*, but not in humans. Explain why this means that quinacrine mustard might be a useful drug to treat sleeping sickness. (AO2) **2 marks**

....

....

....

....

....

13 A student wanted to find the optimum pH of amylase. She pipetted 5 cm^3 of starch suspension into each of five tubes and added two drops of iodine solution to each. She pipetted 2 cm^3 of a buffer solution into each of five different tubes, except that each tube had a different buffer solution and 2 cm^3 of amylase. Next, she put all the tubes into a water bath at 37°C. After 5 min, she added the contents of one of the starch tubes to one of the tubes containing enzyme and found the time taken for the blue-black colour to disappear. She repeated this for each of the other tubes. The table shows her results.

pH	Time taken for iodine solution to become colourless/s
2	970
4	810
6	175
8	147
10	239

a What is a buffer solution? (AO3) **2 marks**

....

....

....

....

b **The student concluded that the optimum pH for this amylase enzyme is pH 8. Is this conclusion correct? Explain your answer. (AO1)** 3 marks

....

....

....

....

....

....

....

c **Suggest a suitable control for this investigation, explaining why it is needed. (AO3)** 2 marks

....

....

....

....

12

Exam-style questions

Read the extract.

Prions are infectious particles that can cause diseases such as vCJD, the human form of BSE. Prions are made of protein. They cause disease by interacting with proteins already present in cells, changing their tertiary structure.

Prions are notoriously difficult to break down and there is a slight risk that surgical instruments might transmit the disease. Using high pressures at 137°C, detergents and UV irradiation have not (5) been successful at destroying prions on contaminated surgical instruments.

Researchers have now found a protease enzyme in a bacterium that lives in hot springs at temperatures of 60–80°C. It is active at pH 12–14 and can break down prions within an hour of incubation. It works by breaking down the bonds between the monomers in the protein.

Use the information in the extract and your own knowledge to answer the following questions.

1 **Suggest why changing the tertiary structure of proteins already present in cells may cause disease (line 3). (AO2)** 3 marks

....

....

....

....

....

....

....

....

2 Most proteins would be destroyed at 137°C (line 5). Explain why. (AO2) **2 marks**

3 a The enzyme that researchers have found is stable at high temperatures and a fairly high pH (line 8). What does this suggest about its structure? (AO2) **3 marks**

b Name:

i the monomers in the protein (line 9) (AO1) **1 mark**

ii the bonds between the monomers in the protein (line 9) (AO1) **1 mark**

Nucleic acids are important information-carrying molecules

Structure of DNA and RNA

DNA is the molecule that carries genetic information in all living cells and RNA is the molecule that transfers this genetic information to the ribosomes during protein synthesis. DNA and RNA are both nucleic acids, which are polymers of nucleotides. Nucleotides contain a five-carbon sugar, a phosphate group and an organic base. In transcription, the DNA in the nucleus is used as a template to make a molecule of mRNA that carries the genetic information to the ribosome. In translation, mRNA attaches to a ribosome and tRNA brings a specific amino acid.

1 Name the three components of a nucleotide. (AO1) **2 marks**

2 How does a DNA nucleotide differ from an RNA nucleotide? (AO1) **2 marks**

3 The diagram shows a section of a DNA molecule.

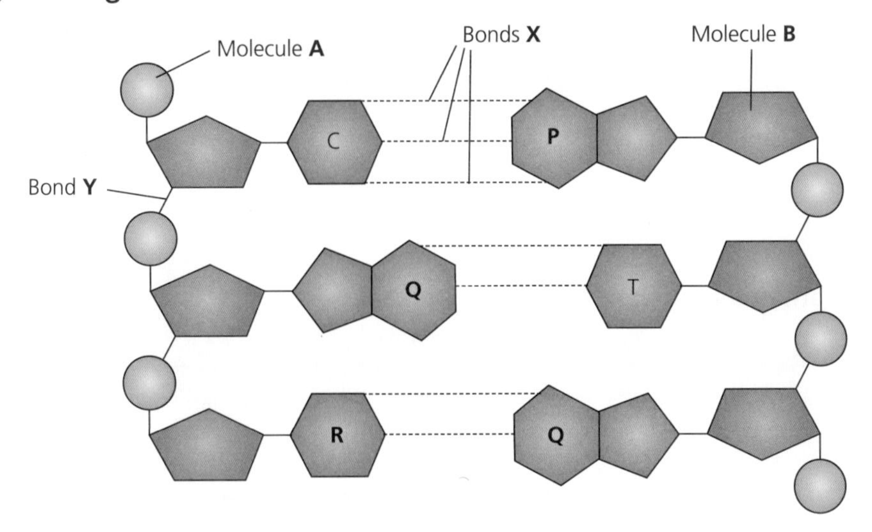

a i Name bonds:

X .... (AO1) 1 mark

Y .... (AO1) 1 mark

ii Name components:

P .... (AO1) 1 mark

Q .... (AO1) 1 mark

R .... (AO1) 1 mark

A .... (AO1) 1 mark

B .... (AO1) 1 mark

b Name the reaction that occurs when bond Y is formed. (AO1) 1 mark

....

4 Give *three* structural differences between DNA and RNA. (AO1) 3 marks

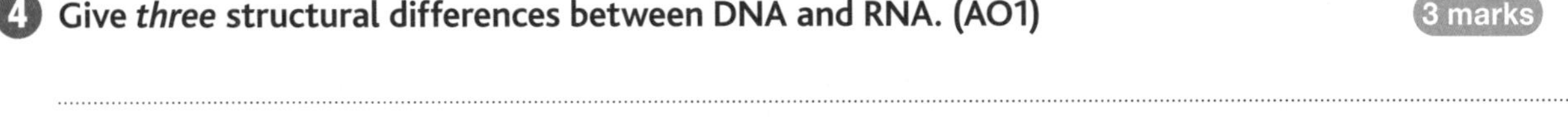

....

....

....

....

....

5 A piece of DNA contains 20% thymine nucleotides. What percentage of the nucleotides contain guanine? (AO1) 1 mark

....

6 Name the *two* molecular components of a ribosome. (AO1) 2 marks

....

DNA replication

DNA replicates itself semi-conservatively, as proposed by Watson and Crick. In this process, the DNA double-helix splits and each strand becomes a template for the formation of a new complementary strand. New nucleotides align alongside the exposed bases by complementary base-pairing. This ensures that DNA is passed on with great accuracy from generation to generation. It can be shown experimentally that DNA replication is semi-conservative, such as with the classic investigation by Meselson and Stahl.

1 Describe the semi-conservative replication of DNA. (AO1) 4 marks

....

....

....

....

....

....

....

....

....

....

2 Describe the role in DNA replication of:

a **DNA helicase (AO1)** 1 mark

....

....

b **DNA polymerase (AO1)** 1 mark

....

....

3 The diagram below shows the sequence of bases on part of one strand of DNA. Complete the diagram to show the bases present on the complementary strand. (AO1) 1 mark

C	G	T	T	A	G	C

Exam-style questions

Two scientists carried out an experiment to find out how DNA replicates. There were two different theories:

- In semi-conservative replication, the DNA double helix splits into two. Each strand becomes a template for the formation of a new complementary strand, so each daughter molecule has one 'old' strand and one 'new' strand.
- In conservative replication, the 'old' DNA molecule stays intact and causes a 'new' molecule to be formed.

To test these theories, the scientists grew a bacterium for many generations in a medium containing the heavy isotope of nitrogen, ^{15}N. This meant that all the bacterial DNA contained heavy nitrogen. Then they transferred the bacteria to a medium containing the normal isotope of nitrogen, ^{14}N. From this point on, all new DNA formed contained the normal, lighter nitrogen atoms.

1 What component of the DNA will contain the labelled nitrogen atoms? (AO2) 

....

2 The scientists extracted DNA from the bacteria when they had been grown for a long time on the heavy nitrogen medium (generation 0) and again after 1, 2 and 3 generations on the normal nitrogen medium. The middle column of the table shows their results.

Generation	Results obtained (semi-conservative replication)	Results that would have been obtained if conservative replication occurred
0	DNA containing all ^{15}N	
1	DNA after one replication in ^{14}N	
2	Results after two generations in ^{14}N	
3	Results after three generations in ^{14}N	

a **These results support the fact that DNA replication is semi-conservative. Explain how. (AO2)** 2 marks

b **In the right-hand column of the table, sketch the results the scientists would have obtained if DNA replication were conservative. (AO2)** 3 marks

ATP

ATP (adenosine triphosphate) is produced in photosynthesis or respiration. Its structure is similar to a nucleotide. When ATP is hydrolysed to adenosine diphosphate and an inorganic phosphate ion, energy is released that can be used for energy-requiring processes in cells. The inorganic phosphate can also be used to phosphorylate other compounds.

1 **Name the *three* different molecules that are found in one molecule of ATP. (AO1)** 1 mark

2 **Give *two* similarities and *two* differences between ATP and an RNA nucleotide.**

a **Similarities (AO1)** 2 marks

b **Differences (AO1)** 2 marks

3 **Describe the roles of:**

a **ATP synthase (AO1)** 2 marks

b ATP hydrolase (AO1) 2 marks

4 What is phosphorylation? (AO1) 2 marks

5 Give *two* circumstances in which ATP is hydrolysed. (AO1) 2 marks

6 Name the processes during which ATP synthesis takes place. (AO1) 1 mark

7 Give *two* uses of ATP in cells. (AO1) 2 marks

Water

Water is an important constituent of cells and it has many properties that are important in biology. It is an important metabolite in some kinds of reaction and also an important solvent. The molecules form hydrogen bonds between them, giving water cohesive properties. It has a high surface tension, a high heat capacity and a fairly large latent heat of vaporisation. All these properties have biological significance.

1 Water is an important metabolite. Name two kinds of reactions in which water is important. (AO1) 1 mark

2 Water is an important solvent. Explain why. (AO1) 1 mark

3 Explain how the following properties of water are biologically important.

a It has a relatively high heat capacity. (AO1) 1 mark

b It has a relatively large latent heat of vaporisation. (AO1) 1 mark

4 There is strong cohesion between water molecules. Give two examples of where this is important biologically. (AO1) 2 marks

Inorganic ions

Inorganic ions are found dissolved in the cytoplasm of cells and in body fluids. Different ions have different roles in cells. The pH of a cell is determined by the concentration of hydrogen ions. Sodium ions have many functions, including the absorption of specific molecules in the small intestine. Phosphate and iron ions are often constituents of biologically important molecules.

1 a Which solution contains a higher concentration of hydrogen ions — an acid solution or an alkaline solution? (AO1) 1 mark

b The concentration of H^+ ions in a solution is 1.4×10^{-5} M. Use the following formula to calculate its pH. (AO2) 2 marks

$$pH = -\log_{10} [H^+]$$

2 Name the type of ion that is important in haemoglobin. (AO1) 1 mark

3 Name *two* molecules that contain phosphate ions. (AO1) 1 mark

4 Give an example of where sodium ions are important biologically. (AO1) 1 mark

Section 2

Cells

Structure of eukaryotic cells

Eukaryotic cells, such as are found in plants and animals, have a nucleus and membrane-bound organelles such as mitochondria, endoplasmic reticulum and chloroplasts. In multicellular organisms, cells become specialised for different functions. Cells in these more complex organisms are organised into tissues, organs and systems.

1 The diagram shows an animal cell. Name the structures labelled A–J. (AO1) 10 marks

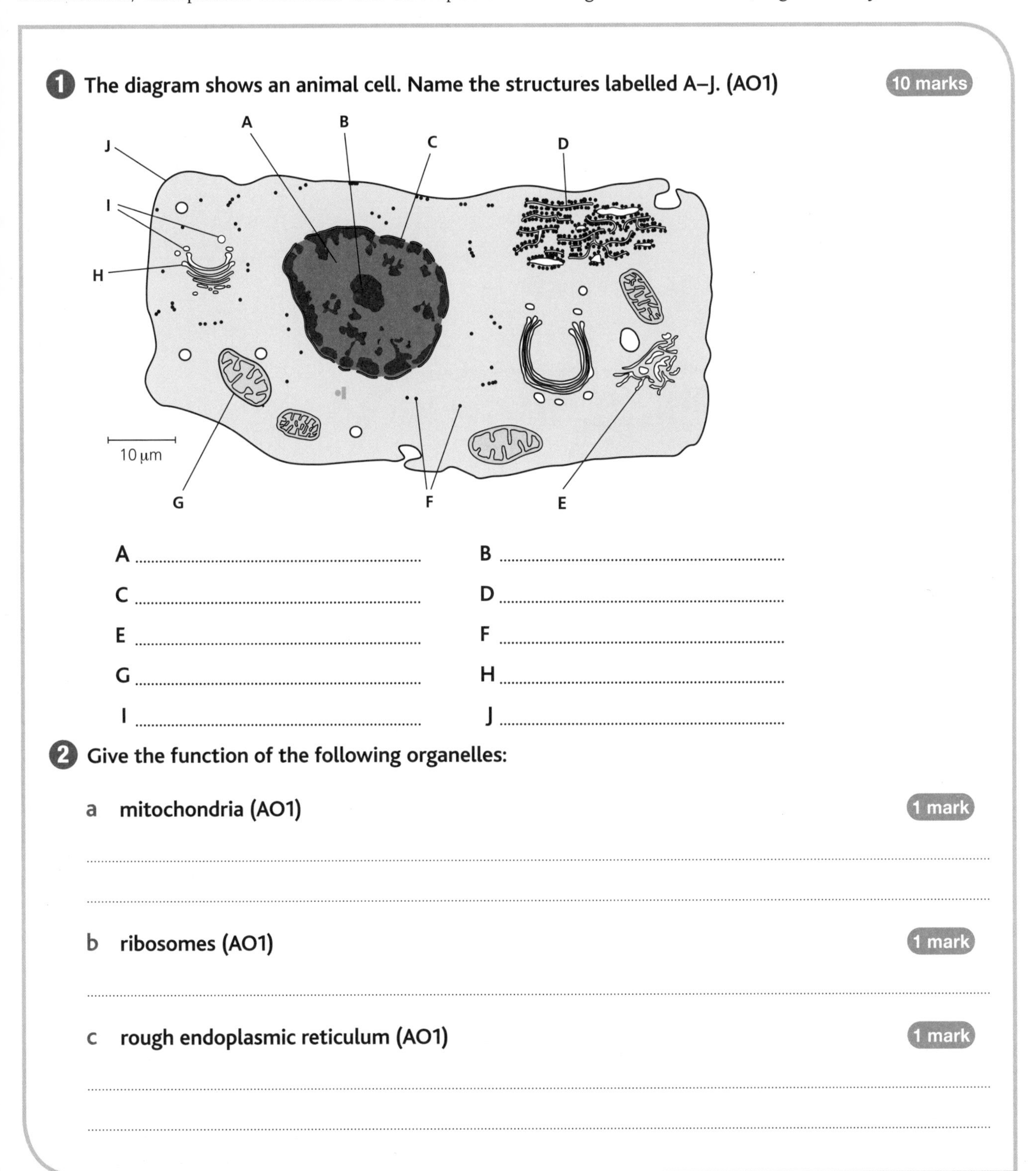

A .... B ....

C .... D ....

E .... F ....

G .... H ....

I .... J ....

2 Give the function of the following organelles:

a **mitochondria (AO1)** 1 mark

....

....

b **ribosomes (AO1)** 1 mark

....

c **rough endoplasmic reticulum (AO1)** 1 mark

....

....

d smooth endoplasmic reticulum (AO1) 1 mark

....

....

e Golgi apparatus (AO1) 1 mark

....

....

f lysosomes (AO1) 1 mark

....

....

3 a Describe the appearance of the DNA in the nucleus of a eukaryotic cell. (AO1) 1 mark

....

....

b What is the function of the nucleolus? (AO1) 1 mark

....

....

4 The diagram shows a plant cell. Name structures A–L. (AO1) 12 marks

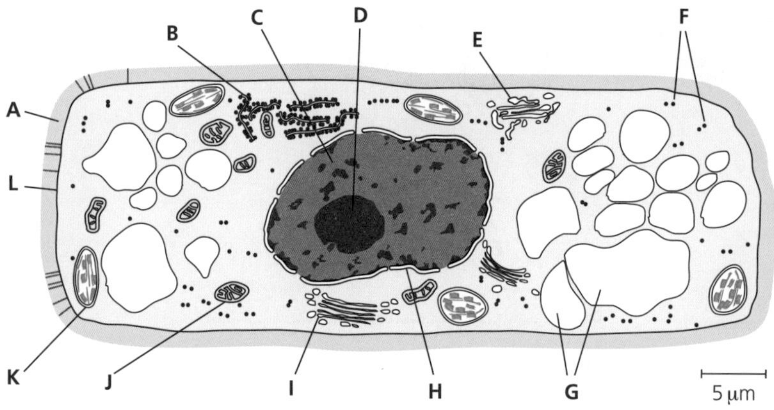

A .... B ....

C .... D ....

E .... F ....

G .... H ....

I .... J ....

K .... L ....

5 a Name *three* structures that are found in a plant cell but not in an animal cell. (AO1) 3 marks

....

b Give the functions of the three structures you named in part a. (AO1) 3 marks

....

....

6 **Why can blood be classified as a tissue? (AO1)** 1 mark

7 **What is an organ? (AO1)** 1 mark

8 **What is a system? (AO1)** 1 mark

Structure of prokaryotic cells and of viruses

Prokaryotic cells are smaller than eukaryotic cells. They do not have a nucleus or membrane-bound organelles. They may contain plasmids and there may be flagella and/or a capsule. Viruses are much smaller than bacteria, which have prokaryotic cells. Viruses are composed of nucleic acid enclosed in a protein coat. They are non-living and acellular. They can replicate only inside a living cell.

1 **The diagram shows a prokaryotic cell.**

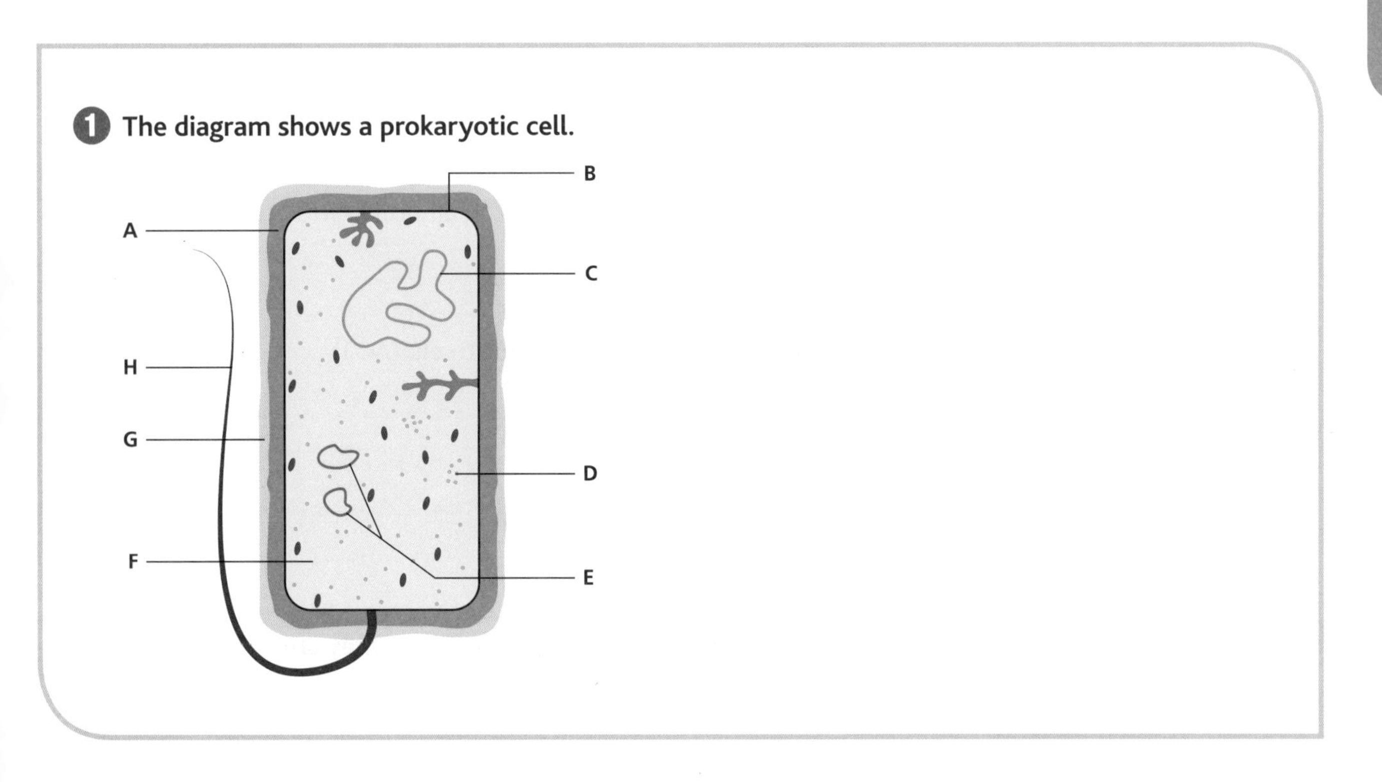

Complete the table below to give the names and functions of structures A–H. (AO1)

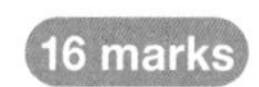

Structure	Name of structure	Function
A		
B		
C		
D		
E		
F		
G		
H		

2 Give *two* ways in which the DNA in a prokaryotic cell is different from the DNA in a eukaryotic cell. (AO1) 2 marks

....

....

....

....

3 Give *two* structures visible in the diagram that are not present in all prokaryotic cells. (AO1) 2 marks

....

4 Give *two* differences between a prokaryotic cell and a eukaryotic cell. (AO1) 2 marks

....

....

....

....

5 Both eukaryotic and prokaryotic cells contain ribosomes. Give *one* way in which eukaryotic and prokaryotic ribosomes are similar and one way in which they are different.

a Similarity (AO1) 1 mark

....

....

b Difference (AO1) 1 mark

6 The diagram shows the structure of a virus.

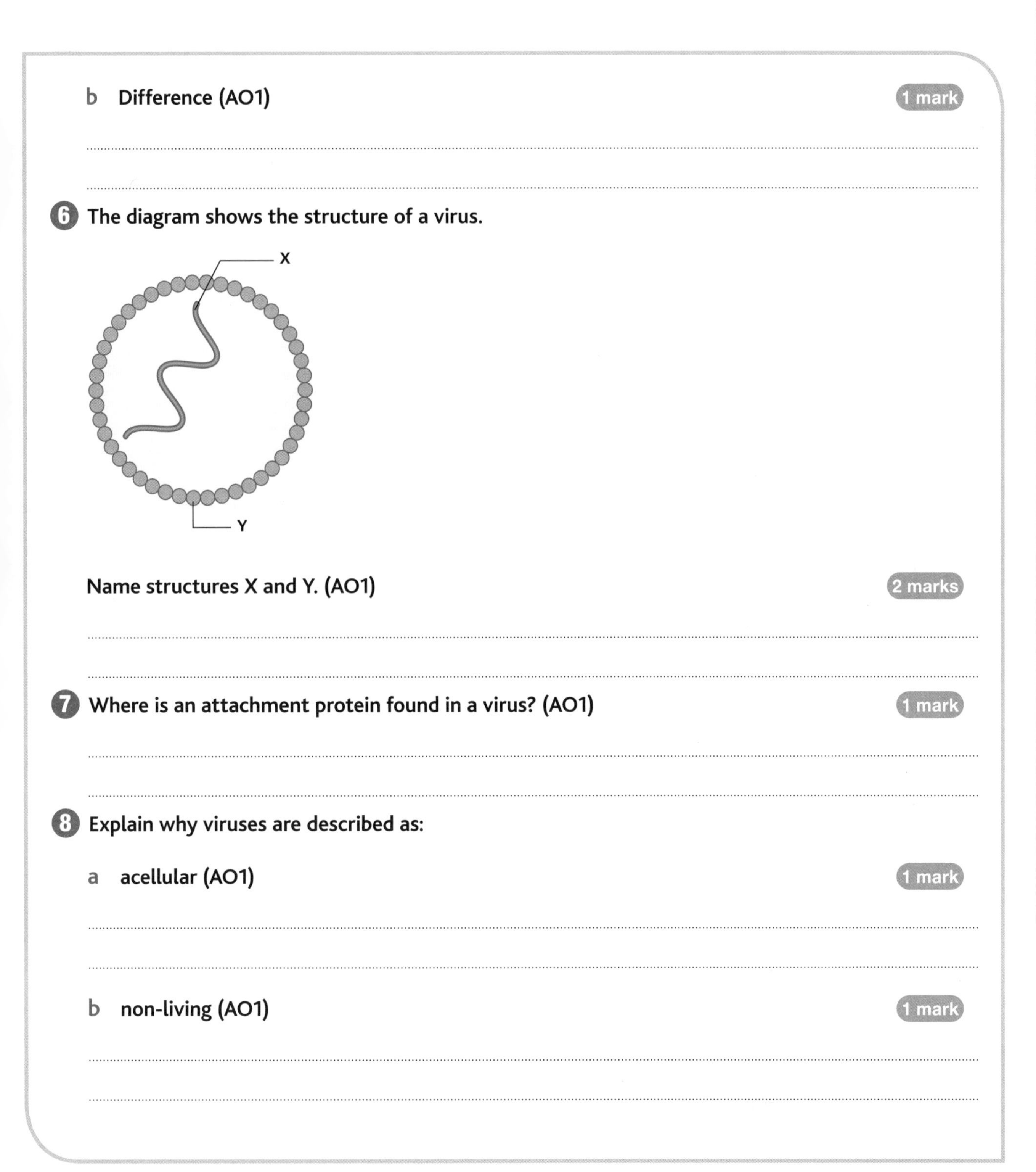

Name structures X and Y. (AO1) 2 marks

7 Where is an attachment protein found in a virus? (AO1) 1 mark

8 Explain why viruses are described as:

a acellular (AO1) 1 mark

b non-living (AO1) 1 mark

Methods of studying cells

Cells can be studied using microscopes. Both electron and optical microscopes are useful in investigating cell structure. Electron microscopes have greater resolution than optical microscopes. The two main kinds of electron microscope are the scanning electron microscope (SEM) and the transmission electron microscope (TEM). If we know the magnification of an image, we can calculate the actual size of structures shown in the image. Cell components can be obtained using cell fractionation and ultracentrifugation.

1 **Explain the difference between magnification and resolution. (AO1)** 2 marks

....

....

....

....

2 **Explain why electron microscopes have a higher resolution than optical microscopes. (AO1)** 1 mark

....

....

3 **Complete the table to describe features of optical and electron microscopes. (AO1)** 4 marks

Feature	Optical microscope	Electron microscope
Image produced	Colours can be seen	
Resolution		Higher resolution
Specimen		Specimens must be dead because they are placed in a vacuum
Ease of use	Portable and easy to use without complex preparation or training	

4 **State the advantages of using:**

a **a scanning electron microscope (AO1)** 1 mark

....

....

b **a transmission electron microscope (AO1)** 1 mark

....

....

5 Calculate the magnification of the cells on pages 23 and 24. (AO2)

6 The real length of the bacterial cell shown on page 25 is 2 μm. Calculate the magnification of this diagram. (AO2)

7 How many nanometres are there in 1 micrometre? (AO1)

8 A structure measures 19 mm in an electron micrograph. The magnification of the electron micrograph is ×1800. What is the actual length of the structure in μm? (AO2)

9 Put these statements in order to describe how cell fractionation and ultracentrifugation can be used to obtain a sample of mitochondria. (AO1) 2 marks

....

A The mixture is filtered through muslin.

B The mixture is spun in an ultracentrifuge at a fast speed.

C Liver from a freshly killed rat is chopped up and made ice cold.

D The supernatant is poured off into a fresh test tube and spun again in an ultracentrifuge at fast speed.

E The mixture is homogenised in ice-cold, isotonic buffer solution.

F The supernatant is poured off and the pellet re-suspended in isotonic buffer solution.

10 Suggest why the liver used to obtain mitochondria in question 9 should be from a freshly killed animal. (AO2) 1 mark

....

....

11 Explain why the tissue used to obtain mitochondria is kept ice cold. (AO1) 1 mark

....

....

12 Explain why the tissue is homogenised in:

a a buffer solution (AO1) 2 marks

....

....

....

....

b an isotonic solution (AO1) 2 marks

....

....

....

....

13 Following the procedure outlined in question 9, what organelle would be present in the first pellet? Explain your answer. (AO1) 2 marks

....

....

....

....

8

Exam-style questions

The electron micrograph shows a mesophyll cell from a leaf.

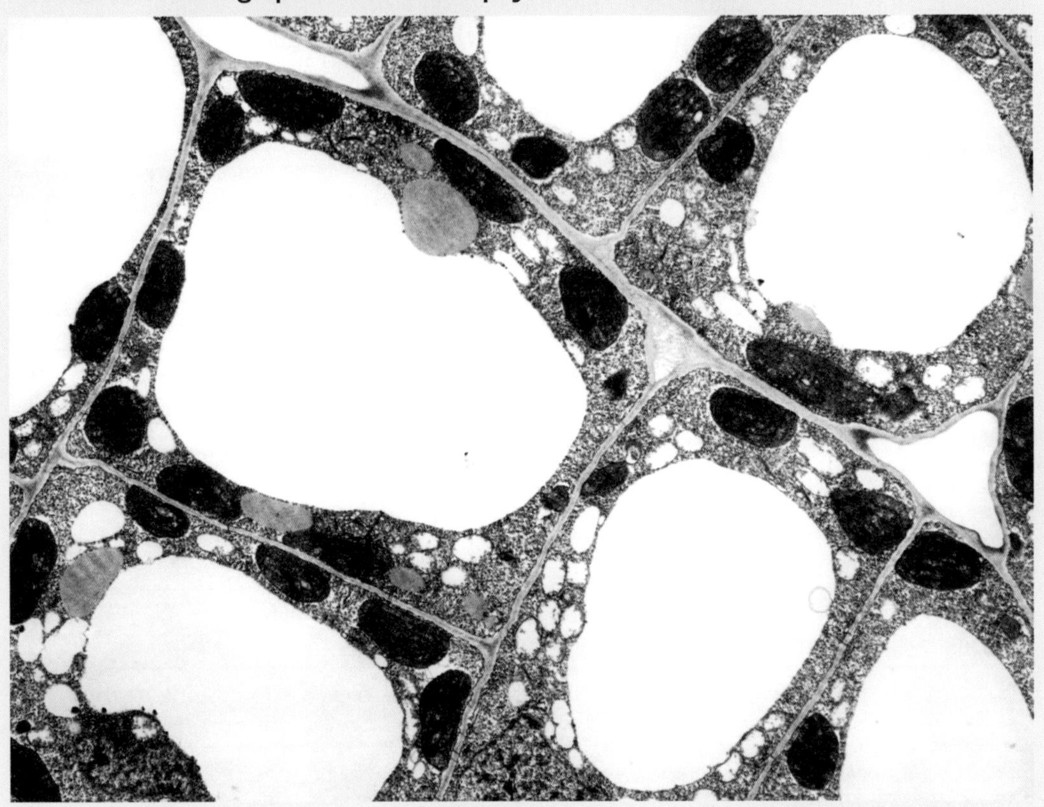

1 Give *two* features visible in the electron micrograph that allow you to identify this:

a as a plant cell and not an animal cell (AO1) **2 marks**

....

....

....

....

b as a eukaryotic cell and not a prokaryotic cell (AO1) **2 marks**

....

....

....

....

2 Which kind of electron microscope was used to produce this micrograph?
Give a reason for your answer. (AO2) 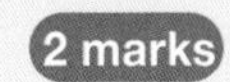

....

....

....

....

All cells arise from other cells

Mitosis takes place during cell division in eukaryotic cells to produce two identical daughter cells. Mitosis is nuclear division and can be divided up into prophase, metaphase, anaphase and telophase. Interphase is the time between nuclear divisions and DNA replication occurs during this time. Cytokinesis is when the cell divides in two. Normally mitosis is controlled, but when it becomes uncontrolled cancer may result. In prokaryotic cells the DNA is replicated and the cells divide by binary fission. Viruses do not undergo cell division because they are acellular and non-living. Instead, viruses are replicated by the host cells they infect.

1 In what stage of the cell cycle does DNA replication occur? (AO1) 1 mark

....

2 Name the stages of the cell cycle described in the table. (AO1) 5 marks

Stage of cell cycle	Description
	The chromosomes line up along the equator of the spindle
	DNA replication occurs
	The centromeres divide and the chromatids (now called daughter chromosomes) move to opposite poles of the cell
	The chromosomes appear. The DNA has replicated so each chromosome consists of a pair of chromatids
	A new nuclear membrane forms around each group of chromosomes

3 Explain the difference between a chromatid and a chromosome. (AO1) 1 mark

....

....

4 The diagrams show a cell in various stages of the cell cycle.

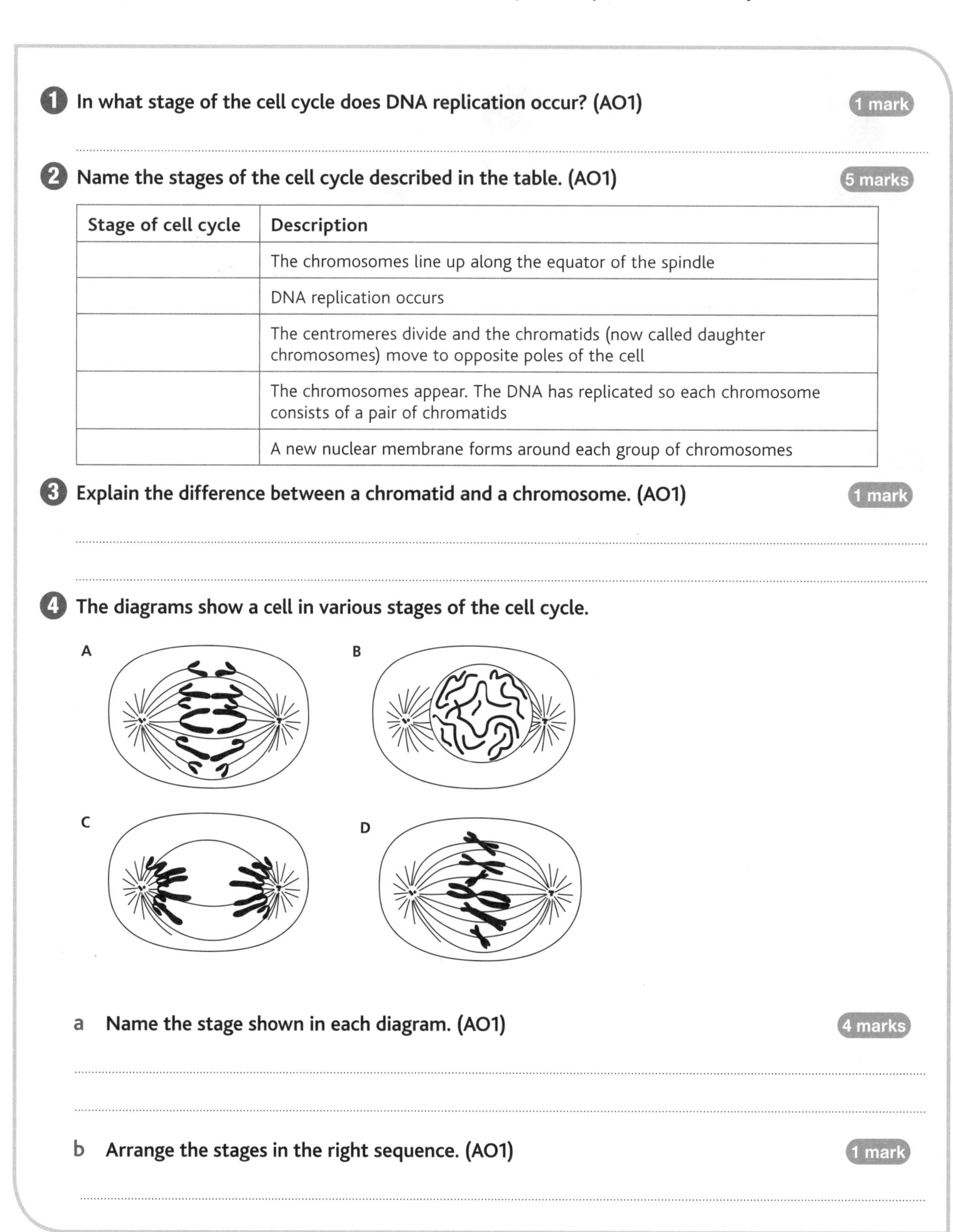

a Name the stage shown in each diagram. (AO1) 4 marks

....

....

b Arrange the stages in the right sequence. (AO1) 1 mark

....

5 The graph shows the changes in the mass of DNA and the total cell mass during two cell cycles. A different scale is used on the *y* axis for each line.

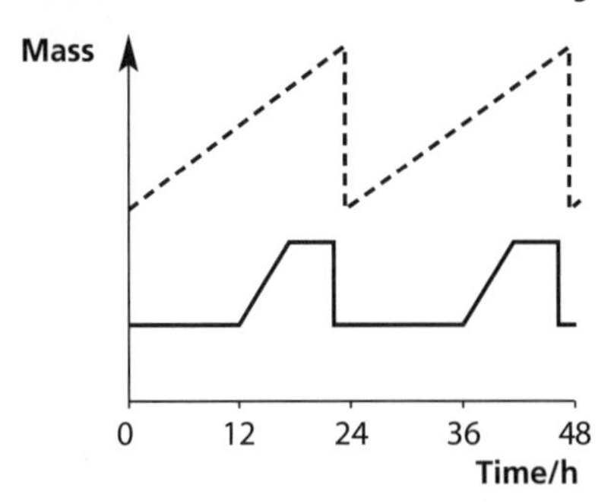

Cell mass - - - - - -

DNA mass ———

On the graph, write the letter:
D to indicate a time at which DNA replication is taking place.
C to indicate a time when cytokinesis is taking place.
M to indicate a time when mitosis is taking place. (AO2) **3 marks**

6 The table shows the number of cells in each stage of the cell cycle in one region of a root tip.

Stage of cell cycle	Number of cells observed
Prophase	16
Metaphase	12
Anaphase	6
Telophase	7
Interphase	159

a Calculate the percentage of time spent in interphase. Show your working. (AO2) 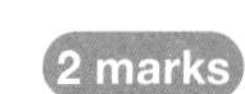

b Calculate the mitotic index for these cells. Show your working. (AO2) 

7 Vincristine is a drug that inhibits spindle formation.

a What effect will this have on mitosis? (AO2) **2 marks**

....

....

....

....

b Vincristine can be used to treat cancer. Explain why. (AO2) **2 marks**

....

....

....

....

8 5-fluorouracil is similar in shape to the base thymine.

a This can be used to treat cancer. Suggest how it works. (AO2) **3 marks**

....

....

....

....

....

....

....

....

b 5-fluorouracil can also harm healthy cells. Explain why. (AO2) **1 mark**

....

....

9 Methotrexate is a drug that inhibits the enzyme dihydrofolate reductase, whose normal substrate is folic acid. The structures of folic acid and methotrexate are shown below.

Folic acid

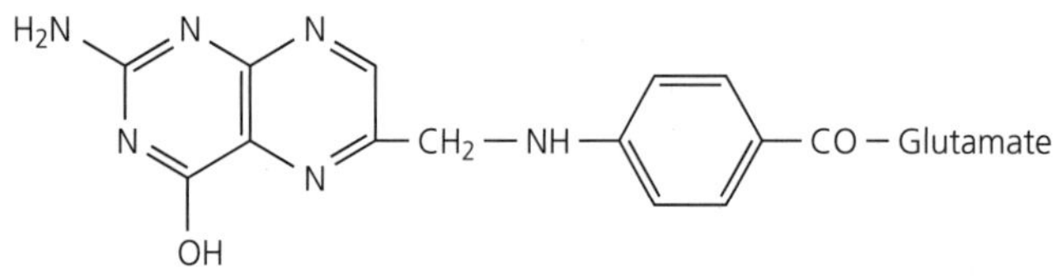

Methotrexate

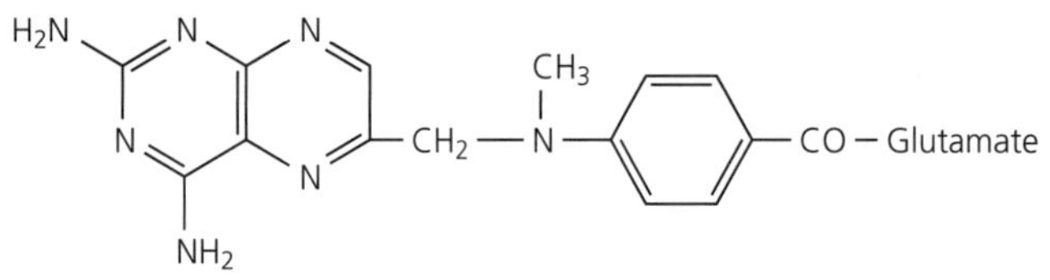

a Suggest how methotrexate inhibits dihydrofolate reductase. (AO2) 3 marks

b When dihydrofolate reductase is inhibited, this prevents nucleotides being produced. Explain how methotrexate is useful as a drug to treat cancer. (AO2) 2 marks

10 The diagram shows cell division in bacteria.

Bacterial DNA
X
Cell replication

a Name the process shown in the diagram. (AO1) 1 mark

b i Name structure X. (AO1) 1 mark

ii Give the function of structure X. (AO1) 1 mark

11 A student cut off about 5 mm from the end of a young growing root. He put this into a glass dish containing dilute hydrochloric acid and acetic orcein stain. He warmed this carefully. Next, he put the root tip into the middle of a microscope slide, added two drops of acetic orcein stain to it and carefully lowered a coverslip on top. He folded a piece of filter paper and placed it carefully on top, then pushed firmly. This squashed the tissue. Finally, the student viewed the slide under an optical microscope and identified which stage of mitosis the cells were in.

a Explain why the student used:

i a root tip (AO3) — 1 mark

....

....

ii dilute hydrochloric acid (AO3) — 1 mark

....

....

iii acetic orcein stain (AO3) — 1 mark

....

....

b Explain why the student squashed the tissue once he had made the slide. (AO3) — 1 mark

....

....

12 Describe the process by which viruses are replicated by host cells. (AO1) — 3 marks

....

....

....

....

....

....

....

....

Transport across cell membranes

All cell membranes, including the cell-surface membrane and the membranes around cell organelles in eukaryotes are the same. Cell membranes are made of a phospholipid bilayer with proteins embedded in it. This is called the fluid mosaic structure. Substances can cross the membrane by simple diffusion, facilitated diffusion, and osmosis, which are passive processes, as well as by active transport and co-transport. Some cells have adaptations for rapid transport across their membrane, such as microvilli to increase their surface area.

1 The diagram shows the fluid mosaic model of membrane structure.

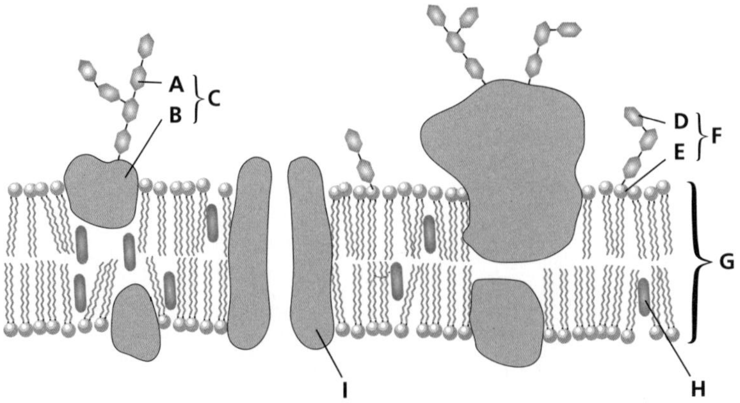

a Name structures A–I. (AO1) **9 marks**

A .... B ....

C .... D ....

E .... F ....

G .... H ....

I ....

b Explain why the membrane is described as:

i fluid (AO1) **1 mark**

....

....

ii mosaic (AO1) **1 mark**

....

....

2 Give *one* function of cholesterol in cell membranes. (AO1) **1 mark**

....

....

3 a What is diffusion? (AO1) **1 mark**

....

....

b Why is diffusion described as passive? (AO1) **1 mark**

....

....

4 Give *one* similarity and *one* difference between simple diffusion and facilitated diffusion.

a Similarity (AO1) **1 mark**

....

....

b Difference (AO1) **1 mark**

....

....

5 Channel and carrier proteins are usually specific to just one molecule. Use your knowledge of protein structure to explain why. (AO1) **2 marks**

....

....

....

....

6 The graph shows the rate of diffusion of two molecules into a cell.

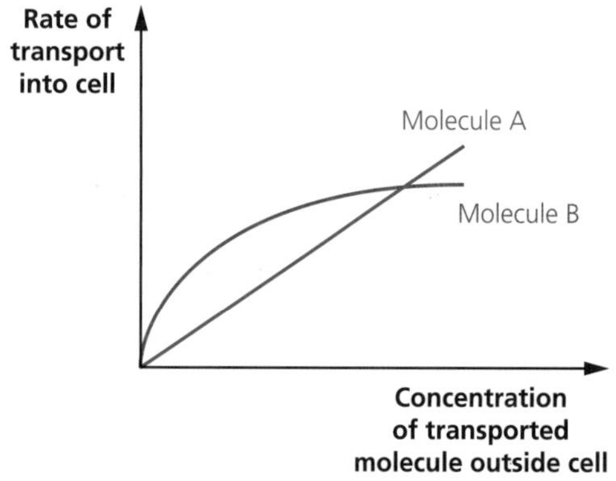

a Describe the graph. (AO2) **2 marks**

....

....

....

....

b Explain how molecules A and B enter the cell. (AO2) **2 marks**

....

....

....

....

7 What is osmosis? (AO1) **2 marks**

....

....

....

8 The diagram shows three cells, each with a different water potential represented by the numbers on the diagram.

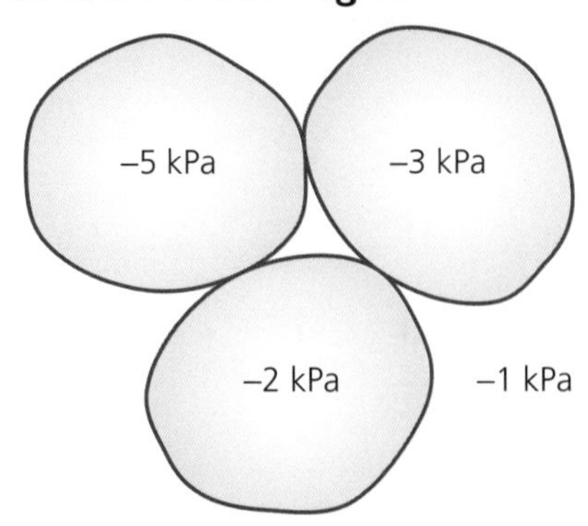

Add arrows to the diagram to show which way water will move between the cells and the surrounding medium. (AO1) **1 mark**

9 A red blood cell bursts when it is placed in distilled water but a plant cell does not. Explain why. (AO1) **2 marks**

....

....

....

....

10 A student wanted to find the concentration of sucrose solution that had the same water potential as potato tissue. She cut cores of potato tissue. She weighed them and placed each one in a tube containing 20 cm^3 of sucrose solution. Each tube had a different concentration of sucrose solution in it. After 24 h she blotted the potato cores dry and reweighed them.

a The student was given a 1.0 $mol\,dm^{-3}$ sucrose solution. Fill in the table to show how she could make up 20 cm^3 of each of the following concentrations. (AO3) 

Concentration of sucrose solution/$mol\,dm^{-3}$	Volume of 1.0 $mol\,dm^{-3}$ sucrose required/cm^3	Volume of distilled water required/cm^3
0.3		
0.6		
0.7		

b Explain how the student could use these results to find the sucrose solution with the same water potential as the potato. (AO3) **4 marks**

....

....

....

....

....

....

....

....

c Was it important that all the potato cores were identical in mass at the start? Explain your answer. (AO3) **1 mark**

....

....

....

....

d The student placed bungs in the tubes before leaving them for 24 h. Explain why. (AO3) **2 marks**

....

....

....

....

11 What is active transport? (AO1) **3 marks**

....

....

....

....

....

....

12 Complete the table with a tick (✓) if the description applies or a cross (✗) if the description does not apply. (AO1) **3 marks**

	Simple diffusion	Facilitated diffusion	Active transport
Uses carrier proteins			
Uses energy from ATP			
Occurs down a concentration gradient			

13 The table shows the concentration of some ions inside the cells of a marine plant and in the seawater in which the plant was growing.

	Ion concentration/mg dm^{-3}		
Sample taken from	Sodium	Chloride	Potassium
Inside the cells	0.11	0.62	0.50
Seawater	0.49	0.58	0.01

How do each of these ions cross the cell membrane? Explain your answer. (AO2) **3 marks**

....

....

....

....

14 Cells lining the small intestine have many microvilli (infoldings of the cell-surface membrane) and contain many mitochondria. Explain how these are adaptations for rapid uptake of substances into the cell. (AO1) **2 marks**

15 Describe how glucose and sodium ions are co-transported into the cells lining the ileum. (AO1) **4 marks**

4

Exam-style questions

The table shows the uptake of a specific ion by a plant cell in different conditions.

Time/min	Ions taken up by plant tissue in different conditions/ arbitrary units		
	Glucose absent, oxygen present	Glucose present, oxygen absent	Both glucose and oxygen present
0	0	0	0
30	0	30	100
60	0	50	150
90	0	70	180
120	0	70	200

1 Describe how you could calculate the rate of uptake of this ion. (AO2) **1 mark**

2 How is this ion taken up by the plant cells? Give reasons for your answer. (AO2) **3 marks**

Cell recognition and the immune system

All cells have specific molecules on their surface that identify the cell. These can act as antigens and enable the immune system to identify pathogens, toxins, abnormal body cells and other foreign cells. These antigens can trigger an immune response. Pathogens are engulfed by phagocytes and destroyed by lysozyme enzymes. T lymphocytes respond to antigens by producing cytotoxic (TC) cells and helper (TH) cells. In response to an antigen, B lymphocytes undergo clonal selection. This results in a clone of plasma cells that produce specific antibodies against the antigen and a clone of memory cells. Memory cells produce a rapid secondary response if the antigen is encountered again. This forms the basis of vaccination. Monoclonal antibodies can be produced, which have numerous uses in medicine and research. The human immunodeficiency virus (HIV) replicates in T cells, causing AIDS.

1 What is an antigen? (AO1) 2 marks

....

....

....

....

....

2 Describe the role of phagocytes in the destruction of pathogens. (AO1) 3 marks

....

....

....

....

....

....

....

3 Describe the response of T lymphocytes to a foreign antigen. (AO1) 5 marks

....

....

....

....

....

....

....

....

....

....

....

....

4 Describe the response of B lymphocytes to a foreign antigen. (AO1) 5 marks

....

....

....

....

....

....

....

....

....

....

....

....

5 What is an antibody? (AO1) 2 marks

....

....

....

....

6 What are monoclonal antibodies? (AO1) 1 mark

....

....

7 How do antibodies lead to the destruction of bacterial cells? (AO1) 3 marks

....

....

....

....

....

....

....

....

8 Explain why a child who has had mumps will not suffer from mumps a second time. (AO1) 2 marks

....

....

....

....

9 What is a vaccine? (AO1) **2 marks**

....

....

....

....

10 The graph shows a primary and secondary immune response.

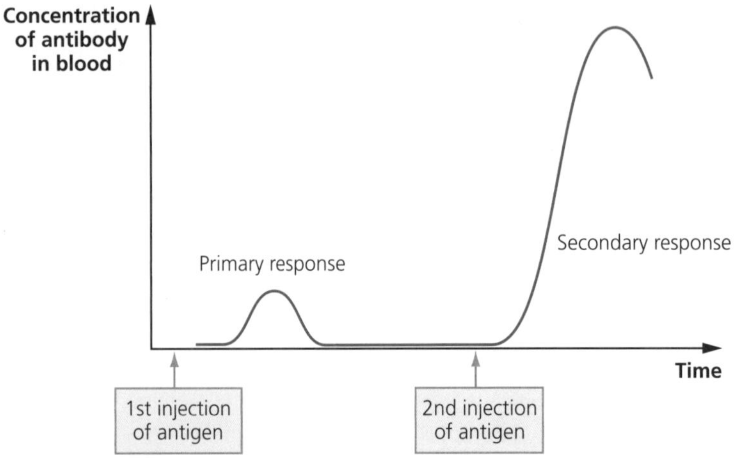

Use the graph to explain how vaccinations can prevent a person from developing certain diseases. (AO1) **3 marks**

....

....

....

....

....

....

11 Explain what is meant by 'herd immunity'. (AO1) **2 marks**

....

....

....

....

12 A drug can be attached to a monoclonal antibody. Explain why this can be particularly useful in the treatment of cancer. (AO1) **2 marks**

....

....

....

....

13 **The chart shows the strains of influenza virus found in one state in the USA during 1 month in the autumn.**

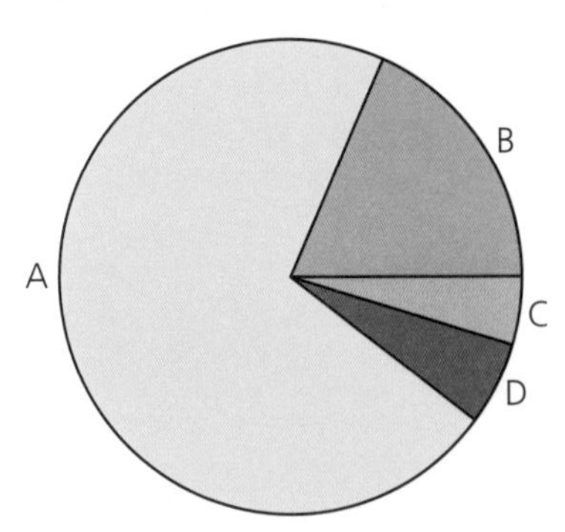

a **Use the chart to explain why people can have influenza on several occasions. (AO2)** 2 marks

....

....

....

....

....

b **Information like this is useful to manufacturers of influenza vaccines. Explain how. (AO2)** 2 marks

....

....

....

....

....

14 a **Explain the difference between active and passive immunity. (AO1)** 

....

....

b **Complete the table with a tick (✓) in the appropriate boxes. (AO1)** 

	Passive immunity	**Active immunity**
A newborn baby receiving anti-bodies in breast milk		
Someone receiving the MMR vaccine		
A fetus receiving antibodies from its mother across the placenta		
Someone being given antivenom following a bite from a poisonous snake		
Someone becoming immune to measles as a result of having the infection		

15 **The flowchart describes how an ELISA test can be used to test for HIV antibodies.**

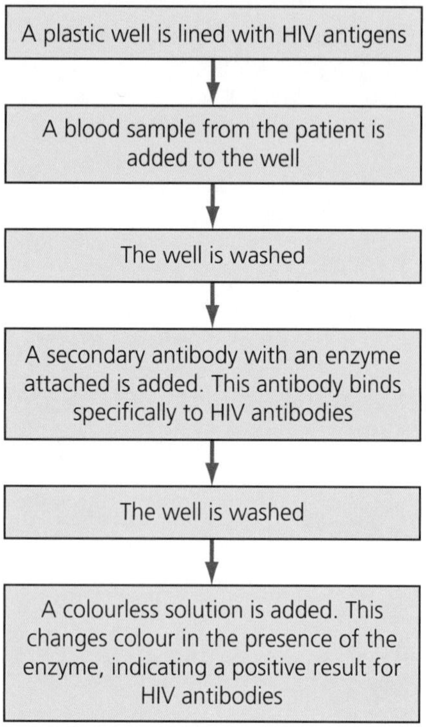

A plastic well is lined with HIV antigens

↓

A blood sample from the patient is added to the well

↓

The well is washed

↓

A secondary antibody with an enzyme attached is added. This antibody binds specifically to HIV antibodies

↓

The well is washed

↓

A colourless solution is added. This changes colour in the presence of the enzyme, indicating a positive result for HIV antibodies

a **Explain why the well is washed at stage 3. (AO1)** 1 mark

....

....

b **Explain why the well is washed at stage 5. (AO1)** 1 mark

....

....

c **Explain the result that would be obtained if the patient's blood does not contain HIV antibodies. (AO1)** 3 marks

....

....

....

....

....

....

....

d **Explain why the colourless solution is added at the final stage. (AO1)** 2 marks

....

....

....

....

16 **HPV is a virus that can cause cervical cancer. It is sexually transmitted. There is a vaccine available that is currently offered to girls in their early teens.**

a **Explain why the vaccine is administered to girls when they are in their early teens. (AO2)** 1 mark

....

....

b **It has been suggested that the vaccination should be offered to boys as well as girls. Evaluate this suggestion. (AO3)** 3 marks

....

....

....

....

....

....

....

....

12

Exam-style questions

Read the extract.

In recent years, crystal meth and ecstasy have become problem drugs. Meth can cause severe problems in the cardiovascular and central nervous systems. In addition, because there is no way to remove the drug from the body, therapies tend to focus on treating its side-effects.

A group of American scientists claims to have developed a way to make antibodies that bind to methamphetamines and similar compounds to effectively remove them from the bloodstream. 5

The team has not yet tested the antibodies in humans, only in rats, but they say that a single injection can reduce the level of drug within the bloodstream for several days.

The antibodies bind to many drugs from the same chemical 'family'. The team says its therapy works for meth, amphetamines and ecstasy, but not for nicotine.

Use the information in the extract and your own knowledge to answer the following questions.

1 **What is an antibody? (AO1)** 2 marks

....

....

....

....

2 **Explain how the antibodies can reduce the level of drug within the bloodstream (line 5). (AO2)** **3 marks**

....

....

....

....

....

....

3 **The antibodies will bind to meth, amphetamines and ecstasy but not nicotine (line 9). Suggest why. (AO2)** **3 marks**

....

....

....

....

....

....

4 **The scientists have tested the antibodies in rats, not humans (line 6). Suggest why. (AO3)** **2 marks**

....

....

....

....

Philip Allan, an imprint of Hodder Education, an Hachette UK company, Blenheim Court, George Street, Banbury, Oxfordshire OX16 5BH

Orders
Bookpoint Ltd, 130 Milton Park, Abingdon, Oxfordshire OX14 4SB
tel: 01235 827827
fax: 01235 400401
e-mail: education@bookpoint.co.uk

Lines are open 9.00 a.m.–5.00 p.m., Monday to Saturday, with a 24-hour message answering service. You can also order through www.hoddereducation.co.uk

ISBN 978-1-4718-4464-5

First printed 2015

Impression number 5 4 3 2 1

Year 2020 2019 2018 2017 2016 2015

Cover photo reproduced by permission of Fotolia

Photo p. 31 © Dr Jeremy Burgess/SPL

Printed in Dubai

Hachette UK's policy is to use papers that are natural, renewable and recyclable products and made from wood grown in sustainable forests. The logging and manufacturing processes are expected to conform to the environmental regulations of the country of origin.